Bibliografic information of the Germans National Library: She registers this publication in the German National Bibliography. Specified bibliografic data are on the Internet under dnb.dnb.de retrievable.

First Edition January, 2024

Erhard Lanzerath

Mühlenberg 14

53902 Bad Münstereifel

Germany

Production and publishing company:

BoD - Books on Demand, Norderstedt, Germany

ISBN: 978-3-384-13547-6

Preface

If one drives through Germany, the numerous, highly rising turbines are very conspicuous. Most were built before 2018. Then the construction tempo decreased drastically. In 2023 has Schleswig - Holstein the biggest wind strength in theirs Arrangements, the Bavarians the slightest one. Arrangements must already become recycled because of her age partial from more than 25 years, the average life expectancy. Their meddlesome size, in many television reports as a great technical achievement in particular with Offshore - Arrangements it is shown, the question provokes: Does it not go also with smaller, more robust and achievement-stronger arrangements?

This book answers this question:

Of course, it does!

The graphics on the pages 24, 26, 27, 28, 29, 31 and 32 are publicly accessible on the Internet and are able to suitable catchwords with Google are found.

Contents

Introduction   1

Small model   3

  Wind machine   5

  Rotor, cutting of a circle, spanning   8

  String role, Umlenkrolle, unit of weight   10

  Achievement control device   12

  Test result energy jump   13

  Test result maximum achievement   15

Wind speed and achievement course   16

Big model   17

  The first test row   18

  The second test row   19

Wind energy   20

Prototype rough draft   21

Facts   23

Appendix   25

Introduction

If one leaves the wind between the long very narrow rotor sheets of customary arrangements unhindered through, so **no subpressure** can form behind the rotor. Therefore it remains whole energy only partially used.

The Leistungsbeiwert, so the proportional use of the wind energy, becomes in the literature for the record simply held in the wind, to the opposition runner, given with 16%. However   If one leaves the wind between the long very narrow rotor sheets of customary arrangements unhindered through, so no subpressure can form behind the rotor. Therefore it remains whole energy only partially used. The Leistungsbeiwert, so the proportional use of the wind energy, becomes in the literature for the record simply held in the wind, to the opposition runner, given with 16%. However is able to the impetus rotor of the same surface to Betz (1920) theoretically up to 59% in energy take up. Modern arrangements reach about 50%. With the ENERCON 126, e.g., (page 32) lies the maximum value with 48.3% and a wind speed of 10 m/s, and they Generator achievement is only half as big like the maximum! This lies since with 17 m/s and an energy exhaustion of only 20%! To the fact, that the ends of the very long wings at enormously high speed by the wind are pressed, statements are absent in the literature. If such ends still really had to go Rotary energy generate? According to the manufacturer GE Renewable Energy they reach Sheet points of the Cypress - 158 speeds from up to 80.3 m/s, this are 289 km/h!

If one wants to reach maximum achievement with a rotor, the wind speed is allowed behind to him amount only 1/3 before him according to Betz. **So the whole rotor surface must be covered with wings.** This is the starting point of the model shown here with the purpose: How broadly if level wings must be and which angle of attack compared with the wind they must have, so that also a considerable **subpressure** is converted into turning force? Becomes with a wind from 15 m/s really a **multiple in energy** compared with 350 watts with the rotors oriented to impetus reaches? And also this question would have to go with a 5 metres - model answerable be: Is the course of the eistungsbeiwertkurve substantially better towards the ENERCON 126 (page 32)? Moreover the surface would have to be bigger with the model under her.

If one tapes the speed and this at a wind measuring place all 10 minutes for a whole year, one can convert the speed into energy, nevertheless, in the end, the **double speed has an eight times higher energy** what often leaves except Eight, if only at middle annual speeds it is calculated. Such an energy curve (page 16) is required, one wants to ascertain from of which wind speed, e.g., 95% of the annual energy are reached. Then for this would have to go the wind power plant are laid out, and it is to be expected that it lies clearly higher than with 15 m/s, required so a very robust rotor which is not throttled possibly already from 12 m/s, how by the impetus arrangements commonly. Just this becomes by the present rotor draught hopefully fulfils.

Small model

With this balsa-wood model (whole view) should be ascertained only, whether the subpressure behind the rotor is into rotary achievement really converted.

Side view
The safety from turn around of the three-legged tower improve striving

Wind machine
"Wind tunnel" (puff)

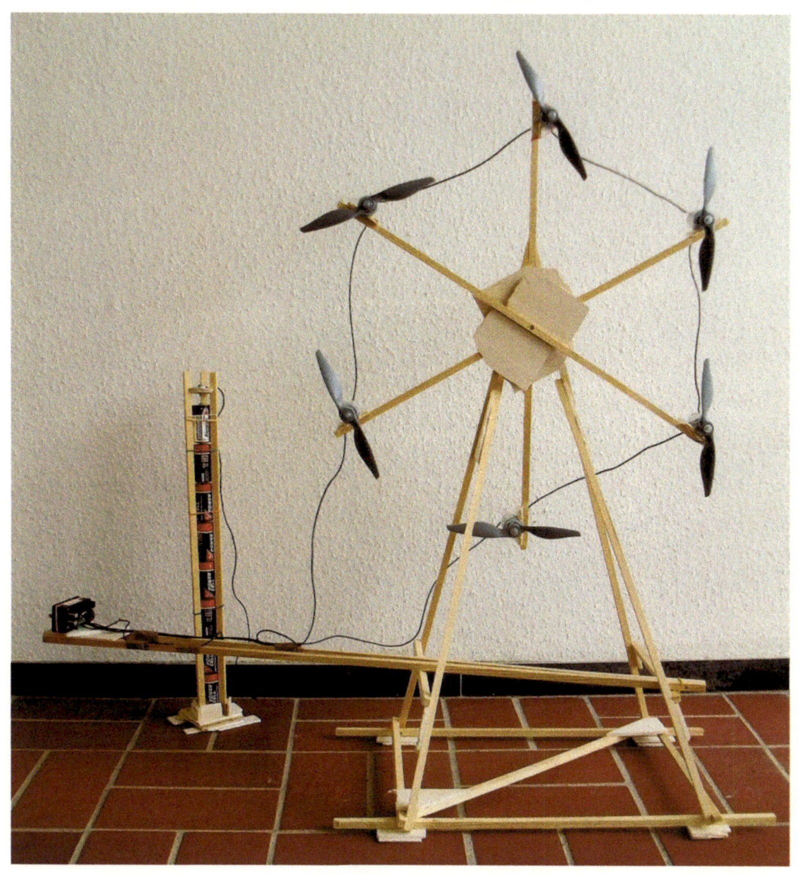

In a row switched model engine. 12 V of battery
tower with electronic achievement control. For
the test all 6 air-screws are put exactly concentric
in front of the rotor, so that the air outside by the
fissures laboriously behind the rotor flows out and
there a subpressure generated.

Model airplane engine

Transparent rotor

The cuttings of a circle with 25 degrees leave visible by the gap with the in parallel running ones edge the air almost unhindered through, but in the test is, that still considerable subpressure behind the rotor into energy is converted (Vollflügler).

Cutting of a circle 60 degrees

The tests have begun with 12 wings 60 degrees of
cutting of a circle. The rotor surface becomes
by the double size with wings covered. Then
everything became gradually about 5 degrees
narrow cut and the arising achievements in a
diagramme held on (side 14).

Rotorspanning

The forwards directed rotor spanning hinders the in each case opposite wings, itself to bend to the back. The corners outside in the wings are likewise braced and prevent the fluttering. Thus a big robustness is achieved.

String role

The string role is connected by the steel axis directly with the rotor. On them becomes long string wound, about a highly hung up role a basket with weights comes up. The height is always 180 cm. Because the time is always same for lifting, so the rotor always equally fast turns, the respective weight is a measure of the required one achievement with different angle of attack of the wings and different width.

Turn around role

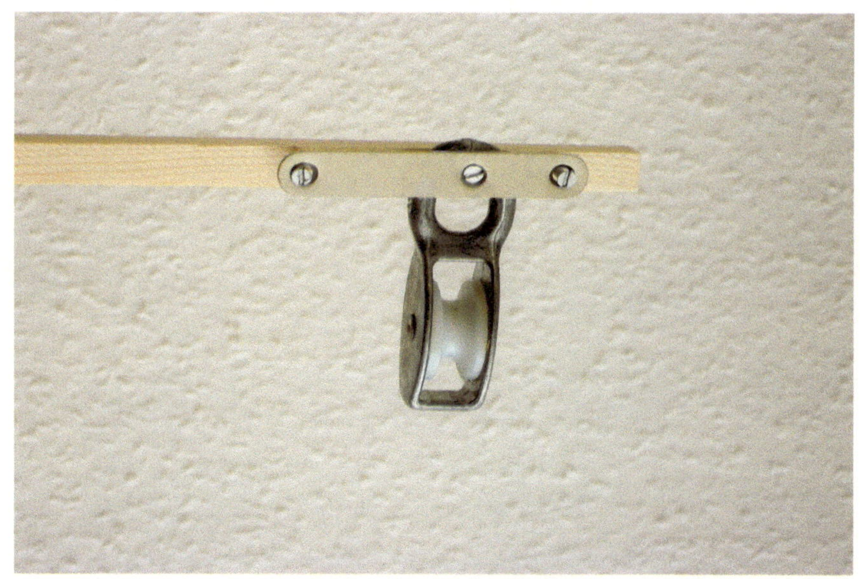

Units of weight

A unit of weight = 1 big and 4 small screws

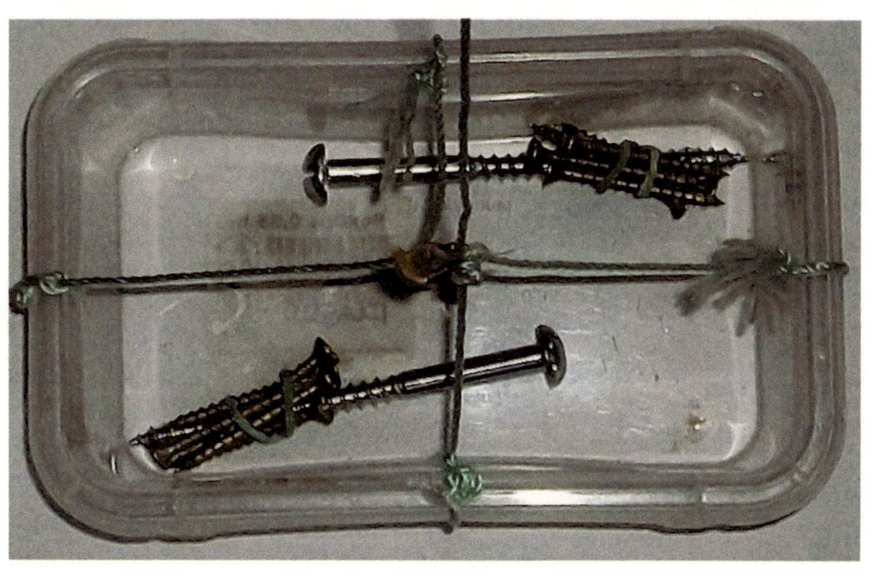

Achievement control device

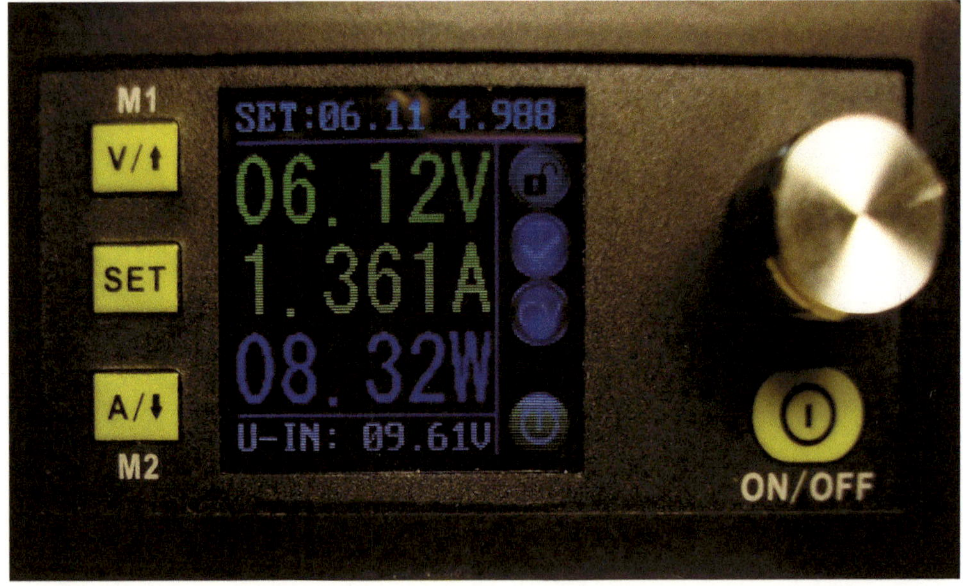

The device permits the precise setting and exact retention of a given achievement in watt - here 8.32 W - even if the voltage of the batteries decreases. For all tests if exactly 8 watts are put. Therefore the same wind energy hits on the rotor always.

Test result energy jump

In the Y - axis page 14 on the left stand the units of weight (UW). In the always same time the weights 180 cm are raised. The angle of attack of the wings is always 45 degrees. If one decreases now the width from 60 to 55 degree of cutting of a circle, it also diminishes Achievement (16UW). Up to 20 degrees the achievement sinks almost straight from on 11.5. From 15 degrees however, the achievement breaks what can only signify, that now the subpressure behind the rotor is not converted into rotary energy any more. By this small model the proof is produced: **The subpressure behind the rotor also becomes with the plain boards in rotary achievement converted**. For the area of 60 to 20 degrees on the Y - axis leaves the energy only about 50% behind. The simple statement from Betz, better than 16% achievement with plain boards does not aplly to neighboring wings and the subpressure behind the rotor.

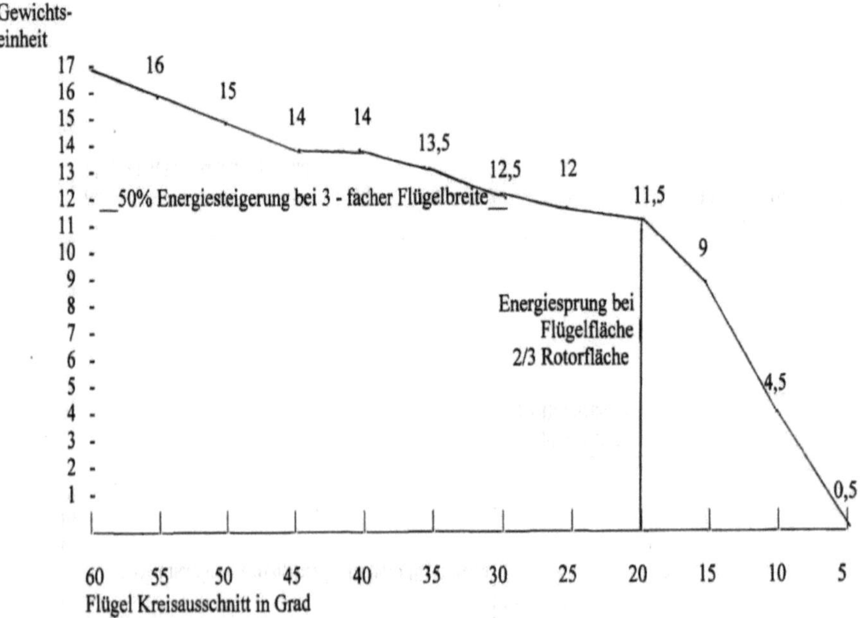

Wind 8 watts, heave 180 cm, angles of attack 45 degrees, the same heave time **Low-pressure loss** at 15, 10 and 5 degrees of cutting of a circle

Test result maximum achievement

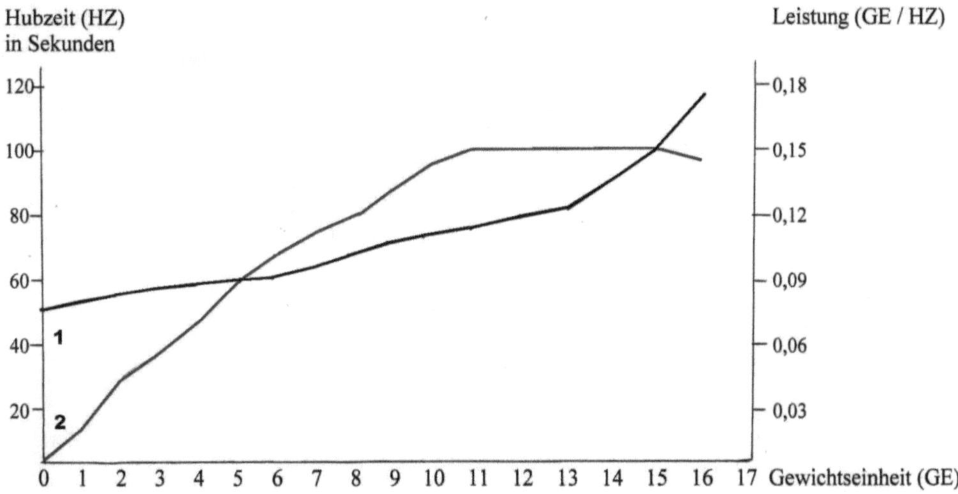

The achievement measurement is expressive a little, because she only for the steady wind energy of 8 Watt and 55 degrees of angle of attack compared with the wind direction counts. In a great model with hence approximative 5 m of rotor diameter extensive tests should take place, which the wind speed, the angle of attack, the wing width and the rotational speeds vary.

1   Heave time (HZ), unit of weight (GE); 2 achievements from GE/HZ; max. achievement 0.15 with 11 - 15 and heave time 100

Wind speed and achievement course

Station ID 90, Alsfeld, 1.1.2000 to 31. 12, 2009 ever 10-minute measurements, commendable way of the DWD provided.

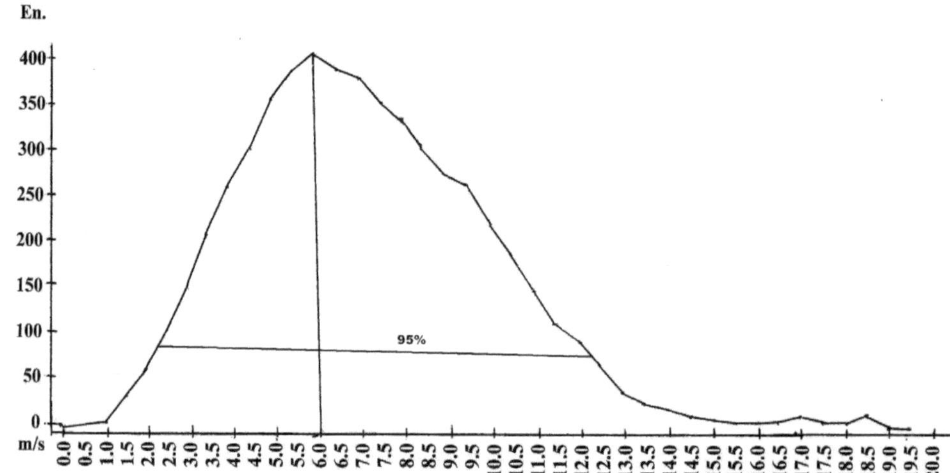

The wind speeds of from 0.0 to 20.0 m/s stand on X - axis. They are on the Y - axis. They are on the Y - axis in energy converted from 0 to 400. 2 m/s energy = 1. So 6.0 m/s in Energy converted are: 6/2 high 3 = 27 energy units (Ee). All measuring values in m/s are groups assigned, thus, e.g., 6.0; 6.1; 6.2; 6.3 and 6.4 in the group 6.0. The suitable Ee are with 6.0 added up and show the wind achievement in this group in the course of 10 years, above the perpendicular 394961. 400 on that Y - axis on top stand for 1000 Ee. The whole 10 - annual achievement is 5408748 Ee. If someone wants to use the energy for 95%, this lies in the area of 2.3 m/s and 12.3 m/s. One maximum exhaustion of this wind energy assumes, that in the whole wind area of 2.3 m/s to 12.3 m/s the efficiency of a wind - turbine with approximate 50% lies. The ENERCON 126 (page 37) however, has with 9 m/s her biggest efficiency with 48.3%, however, the generator performs on this occasion half power. A by far **steadier achievement** use promises the **new rotor construction**, around it goes here!

Big model

With the small model could be proved only, that to itself a low-pressure area considerably has a great effect to the power production. Now the purpose of a big test model lies in that Question: How many watts one is able to win from 1 square metre of rotor surface with a wind speed from 16 m/s? The tests would have to come to the result, that a multiple of electric energy is available compared with approximate 350 watts per square metre of the large turbines. For this the rotor has to go really robustly and must be constructed by easiest technology. Automatic pitching of the wings are useless. Pitching may be done by hand. The rotor diameter of 5 m does not provide for the fact, that whirl in the rotor edge considerably hinder the subpressure. The maximum achievement would be allowed to do with 16 m of wind hardly 40 kW overcome what the generator must be able to take.

The first test row

It is carried out with 16 m/s wind speed. The wings have 60 degrees cutting of a circle and are able to in 7 positions by hand compared with the wind direction opposed become: From 35 degrees in steps of 5 degrees to 65 degrees. It is tested, with which settings and with which rotational speed the maximum achievement are reached.

Achievement per square metre
is important. From her arises the
Achievement factor
as a multiple of 350 watts by the large turbines. If it ist clearly larger than 3, other tests take place. The achievement in watt points, which energy can gained by this rotor - tchnique from the wind **practically**. Thuch a value is nowhere published.

Also the course of the curve for the "Leistungsbeiwert" is established and compared with the ENERCON 126 (Page 32). She shoud not drop sharply in the upper part at the right and left side of the curve, but clearly level one run, because the wings do not become pitched and the rotor surface is covered. This should be a clear advantage by the new rotor - technic, particularly as he by the robust **construction** method also the wind energy more than 16 m/s at most in stream moves. It is not throttled, how by the large turbines.

If it has turned out that it is worthwhile to develop a prototype, next investigations are necessary. It serves only the development of a wind strength arrangement ready to go into mass production and has the qualities: Stationary wings and assembly on a three-legged steel grid tower.

The second test row

It changes the width with the big model. The impractical cutting of a circle of 60 degrees becomes reduced in steps of 5 degrees down to 5 degrees. The angle of attack to the wind direction reaches from 35 to 65 degrees and the wind speeds between 2 and 20 m/s. Therefore the spanning must be performed very robustly. Numerous measuring rows results are to be shown as graphics. The purpose is to find out, with which width the energy admission nearly is optimum what can be settled with the wind achievement graphic arts page 16. Approximately 95% of the annual energy should be exhausted by the rotor if the angle of attack and the width are fixed. Then with this knowledge the prototype may be contructed up to a mass production.

Wind energy

Wrong results of the wind energy follow from a middlesized Wind speed. Moreover a model consideration, from two measuring values in hourly distance goes out:
    1 hour 7 m/s   1 hour 13 m/s
The average of the speeds in m/s is:
$$7 + 13 = 20 / 2 = 10$$
Now 7 m/s = 1 as an energy unit. The energy of 2 hours ist calculated in this way (double speed = 8-fold energy!):
$$(10 / 7) \text{ high } 3 = 2.96 * 2 = \mathbf{5.832}$$
However, the actual energy is for the 2-nd hour:
$$(13 / 7) \text{ high } 3 = \mathbf{6.404}$$
In addition, for the first hour:
$$6.404 + 1 = \mathbf{7.404}$$
One sees averages for the energy calculation in the hourly or even bigger distances, lead to wrong results. Only very short time intervals, e.g., from 10 minutes, lie energy values so near together that they hardly differ of each other. This must considered at all energetic statements. If one looks the ENERCON 126 at (page 32), the following arises:

In the first hour with 7 m/s wind ENERCON works with maximum energy-value of 0.48, but in the subsequent hour of richest energy only with a value of 0.35! It appears, that a steadily wide energy exploitation about a big area of the wind speed from e. g. 3 to 17 m/s is advantageous. The present rotor tchnology promises to own exactly these qualities.

Prototype rough draft

The most important quality for the complete use of the wind energy in the area of about 3 to 17 m/s is the robustness of the rotor. With not turning wings (1). They run on a big steel ring (2) on roles (5):

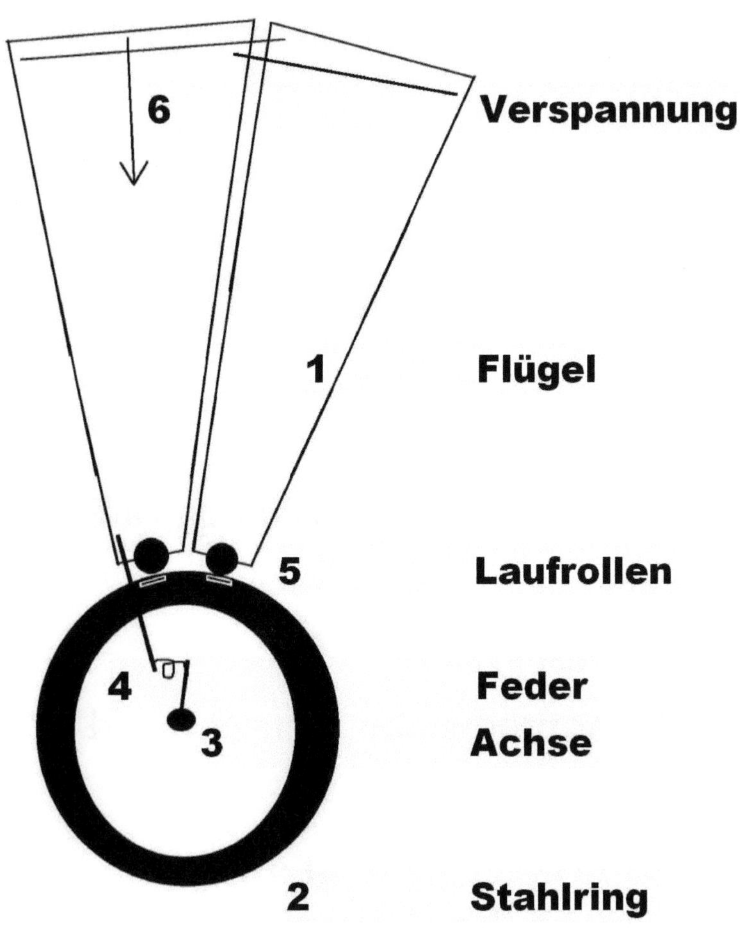

The wings about a feather (4) on the axis (3) transfer the turning energy. Other forces are transfered from the steel ring to the gondola and not also on the axis. The wings get stability by her spanning (6). The angle of attack of the firmly mounted wings towards the wind and the rotational speed has to correspond to the maximum exhaustion of the wind energy in the area e.g. from 3 to 17 m/s. If the wind exceeds the generator achievement, the wheel is turned about approximately 100 degrees from the storm by which it is blown from the back and it comes to a shutdown. Now it is decoupled and the freely running rotor without load is turned into the storm turned back where he turns unloaded cheerfully fast.

In the disclosure writing of the patent

DE 10 IN 2023 000 172.4

details are included.

The advantages of the new rotor technique are:

- Multiple power

- Compact and robust system

- Very low costs for production, transport and
  operation

Let us begin this project straight away!
Then we will manage the „Zeitenwende" (Word of the German Chancellor)

Facts

Often, but also frivolously overlooked facts:
1. The energy of the wind increases with the third power of the speeds. So: double wind speed = 8-fold energy! All the other power stations need for double achievement also the double surface, **as for example with the solar energy.** The German Weather Service collects values the wind speeds concerned and not the wind energy.
2. Coal-fired power stations generate stream to the slightest costs. Other power station kinds are clearly expensive, the WGs enclosed. The solar energy ist most expensive, see page 31.

3. The greater the WGs become, the more slightly the achievement is in watt per square metre. Between 2000 and 2019 she sinks from 410 watts to 340 watts ("Volllaststunden von WEA" of German WindGuard).
4. Most energy gets lost with the lift-based "giants" by the fact, that the Wind everywhere between the wings unhindered streams through, what is not the case with the present rotor construction.
5. A higher hub generates more energy? **No!**

The wind measurement lasted three days from the

4th to the 7.1.2024 in the same mast by the heights from 33m to 100 m. 8 curves coloured for this lie on top of each other or extremely near together, so that energetically seen the hub height seems insignificant, particularly as WGs possibly throttle (pitch) from 10 m/s and the energy in the small area above 14 m/s is not used completely.

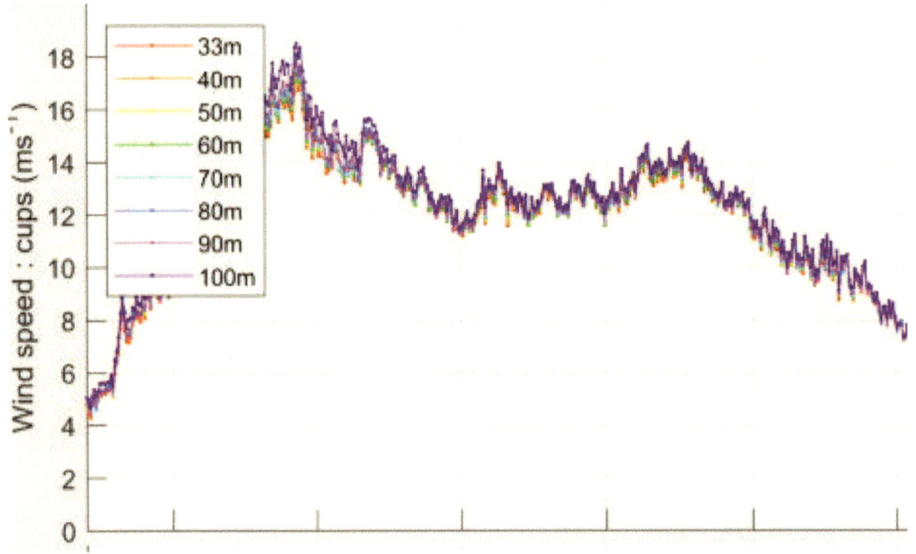

Appendix

Survey 1: CO2 and temperature 160000 years
            ago until today   26

Survey 2: The global warming 1850 to 2020   27

Survey 3: Wind stream today   28

Survey 4: The stream net of the federal net -
            agency   30

Survey 5: Stream production costs, kind of power
            station and kWh   31

ENERCON 126 Leistungsbeiwert   32

Survey 1: CO2 and temperature 160000 years ago until today

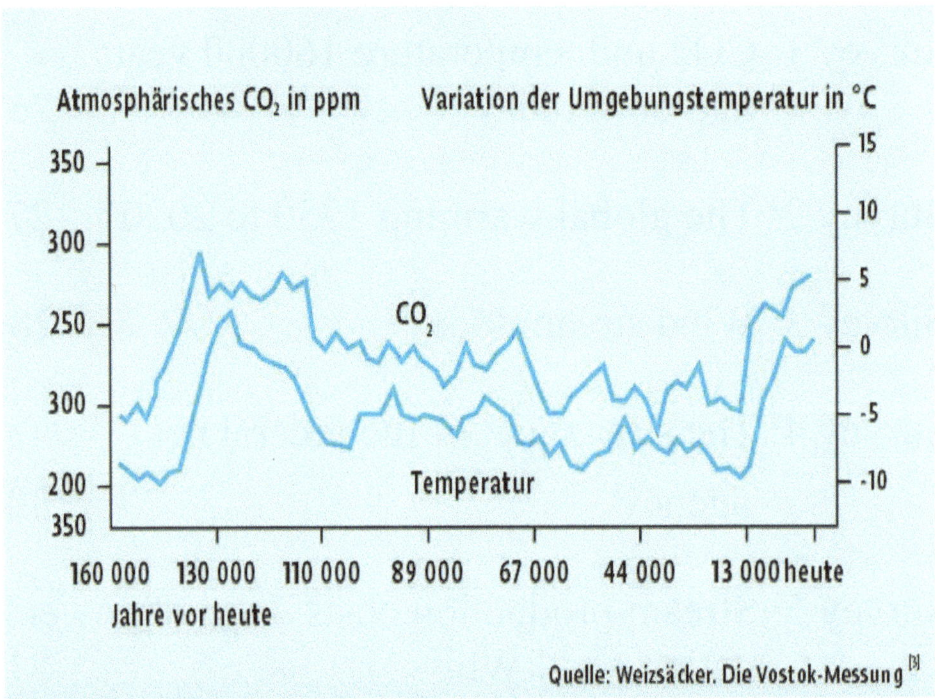

- CO2 maximum 135000 years ago like today
- Temperature maximum 5000 years later
- natural fading away of the CO2 - salary in 90000
  years
- clear decline of the temperature in 10000 years
- no appreciable influence of human being
  A drastic restriction of the additional CO2-
  Load does not lead to the dismantling in the air
  and has also not the immediate decline of the
  temperature as the result. **The earth needs long
  time, to correct our mistakes.**

Survey 2: The global warming 1850 to 2020

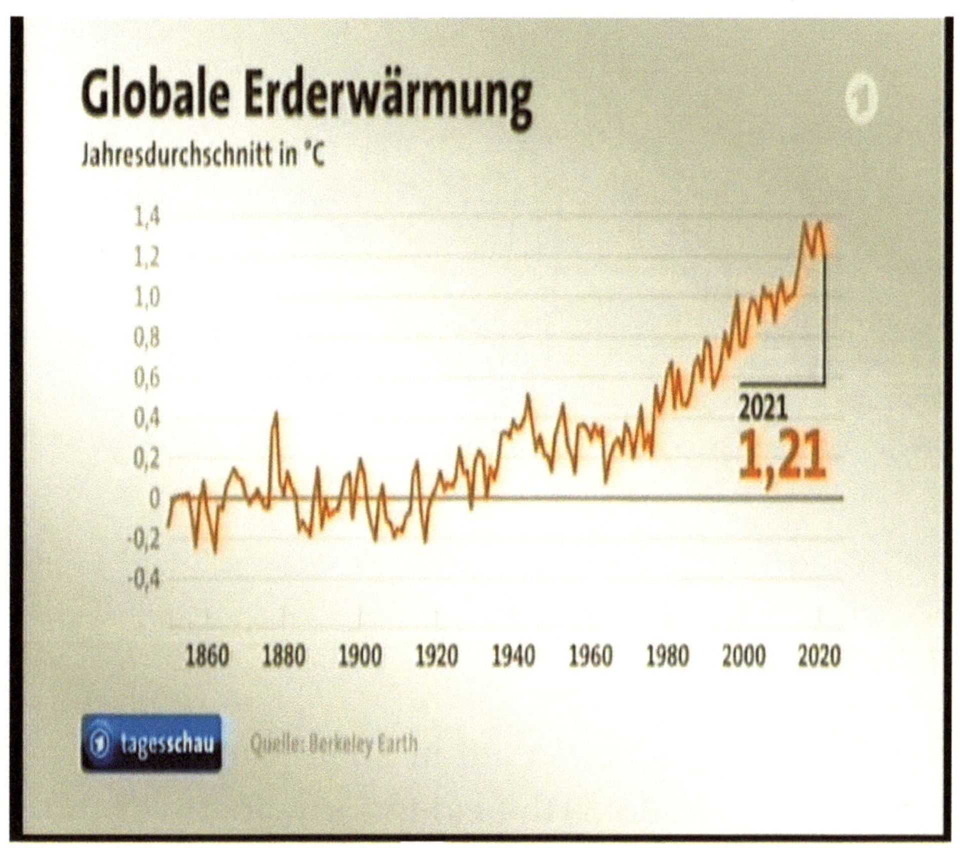

- Small ice age before 1800
- Increase begins with the industrialisation, cutting
  forests and the wide agriculture founds with the
growth of her Humanity after the motto: Power to
you them Earth subject!

Survey 3: Wind stream today

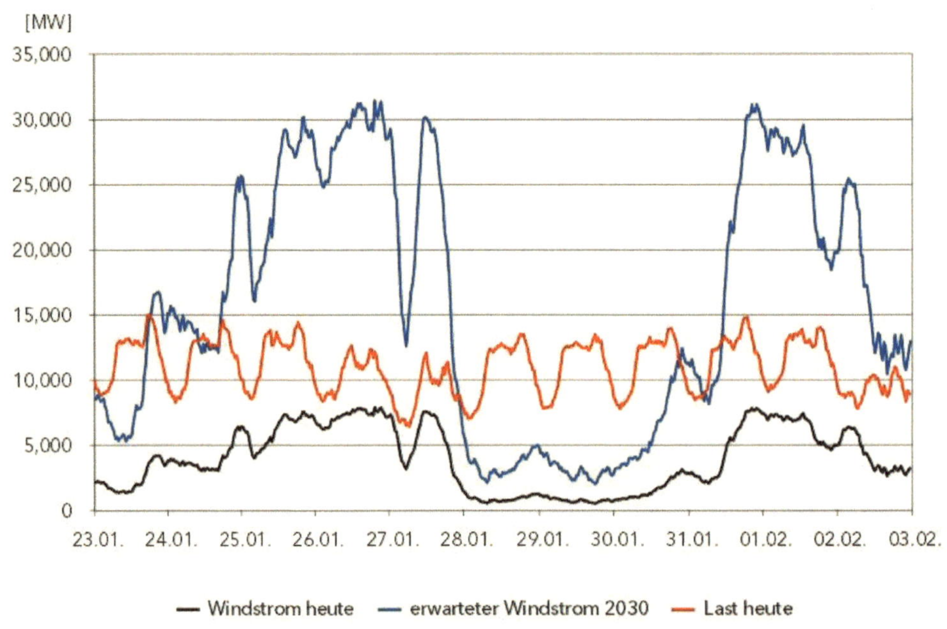

-The wind stream expected in 2030 (blue ones curve) is at 28th to 31th also too low to cover the load from 8 000 to 13 000 kW.

- Numerous power consumers also provide at night for the fact, that at least 8 000 kW are needed.

- Hydrogen power stations would have to go with wind lull to generate the necessary stream. In Prenzlau the hybrid power station ENERTRAG

works since 2011. It generates direct stream from
WGs as well as of self-generated hydrogen.
- The red curve is able to do itself clearly in
future by the electric cars change.
- The green stream generated today from wind,
water power and solar arrangements lies with
about 15% and conventionally generated with
about 85%. Quite different results points ntv

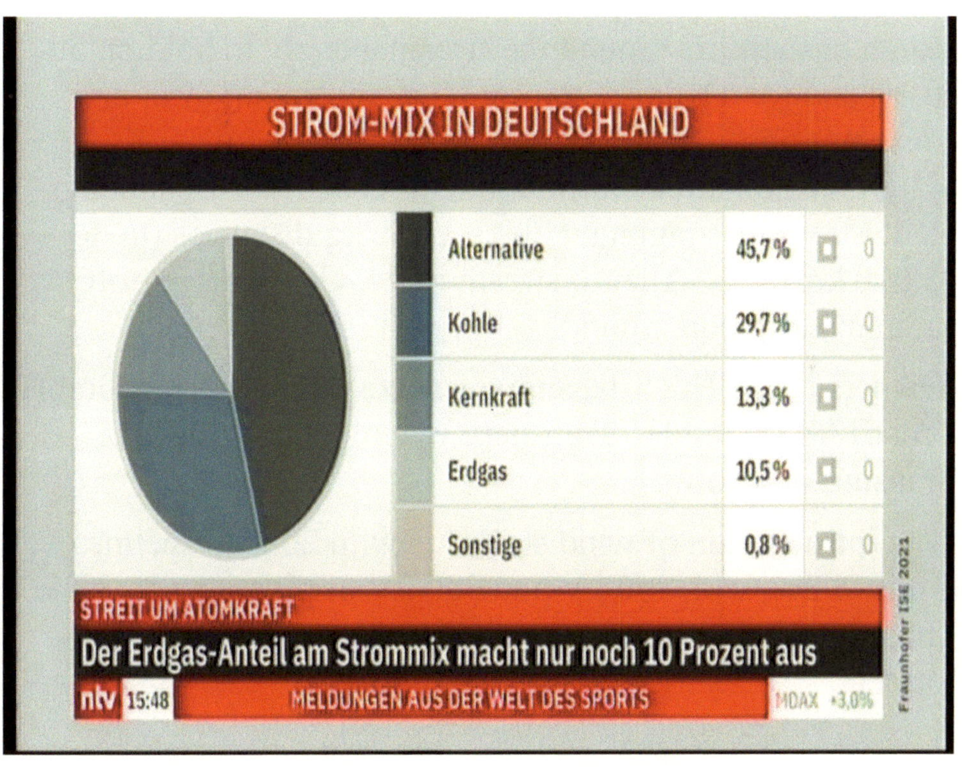

Survey 4: The stream net of the federal net - agency

- Verteilnetz with 230 volts ac stream to household
- Medium tension net with 20,000 volts some km to 100 km
- High-voltage transmission system (more than 60,000 volts) 110000 Volt distributes in regions, areas of concentration and to industrial locations
- Höchstspannungsnetz 380,000 volts away distances. connection also abroad, connection of big power stations to 300 km. One Übertragungsnetzbetreiber near Cologne steers for completely Europe to and switching off from Höchst-spannungsnetzen, around the stream exactly to hold on 50 hearts and for a balance to provide from stream offer and consumption.
- High-tension direct current transference (Hochspannungs-gleichstromübertragung, HGÜ) very big distances (up to 35,000 km), e.g. Allegro stream route. Converters stroll again in ac stream around.

-Owners of the Höchstspannungsnetze are: Amprion GmbH, TenneT TSO GmbH, TransnetBW GmbH und 50Hertz Transmission GmbH

-The introduction of wind stream is with 20 kV Pipelines limits to 2%, with 110 kV and 380 kV of pipelines on 20%. Then the wind stream is stretched to high over own trans-formers.

Survey 5: Stream production costs, kind of power station and kWh in Germany

Kernenergie	3,5 Ct.
Braunkohle	2,8 Ct.
Steinkohle	3,3 Ct.
Gas	4,2 Ct.
Wasserkraft	10,2 Ct.
Wind (on-shore)	7,6 – 12,7 Ct.
Wind (off-shore)	10,0 – 16,1 Ct.
Biomasse	9,6 Ct.
Photovoltaik	50 – 60 Ct.
Solarthermie	10 – 40 Ct.

ENERCON 126 LEISTUNGSBEIWERT

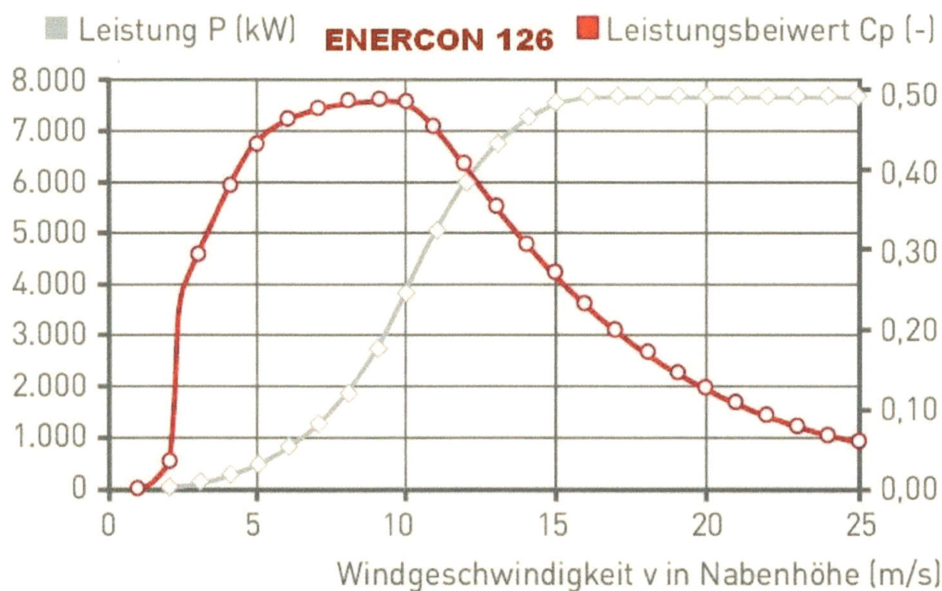

Left   Y - axis   Power in Kolowatt

    X - axis   Wind speed in meter per second

Right Y - axis   Leistungsbeiwert

Erhard Lanzerath became the relay and tubes calculator at the Bundesfachschule für Datenverarbeitung as well as in the Gesellschaft für Mathematik und Datenverarbeitung (GMD) qualified. He was an adviser in the Gesellschaft für Elektronische Informationssysteme (GEI). During many years he developed computer boards and wrote for them operating systems. His advice was asked also in Siemens and Dresdner Bank for great computers. As more independently systems analyst Ho took also a large-scale project after several years' activity with the military. He accompanied his activities with own piano play and song and has often turned to projects, in which he could win new knowledge by radical analysis. Now this also counts to the new one Rotor technology by wind strength arrangements. His wish is, that possibly many people this Technology try out.